'rom:
book
book
Teacher's Manual
Lesson Plan
Study Notebook
Quizzes
Graphics Package

FOCUS ON ELEMENTARY

3rd Edition

Preview Booklet

Rebecca W. Keller, PhD

Real Science-4-Kids

Illustrations: Janet Moneymaker

Focus On Elementary Biology 3rd Edition Preview Booklet

Published by Gravitas Publications Inc.
www.gravitaspublications.com
www.realscience4kids.com

Introduction

Welcome to the ***Focus On Elementary Biology 3rd Edition Preview Booklet*** where you can take our one semester unit study program for a test run!

The materials sampled in this book are taken from a full semester course, with one chapter from each part of the curriculum:

- The ***Focus On Elementary Biology 3rd Edition Student Textbook*** provides foundational science concepts presented in a way that makes it easy for students to read and understand. The many colorful illustrations make each chapter fun to look at and reinforce concepts presented.
- With two science experiments for each chapter, the ***Laboratory Notebook*** helps young students learn how to make good observations, an important part of doing science. Open-ended questions help students think about what they are learning, and information is provided to assist students with understanding what they observed while performing their experiments.
- The ***Teacher's Manual*** includes instructions for helping students conduct the experiments, as well as questions for guiding open inquiry. The commonly available, inexpensive materials used for all the experiments can be seen in the complete materials lists included in this booklet.
- Using the ***Lesson Plan*** makes it easy to keep track of daily teaching tasks. A page for each chapter in the ***Student Textbook*** has the objectives of the lesson and questions for further study that connect science with other areas of knowledge, such as history; philosophy; art, music, and math; technology; and language. Forms are included for students to use to do a review of material they've learned and to make up their own test for the chapter. Also included are icons that can be copied onto sticker sheets and used to help plan each day of the week.
- Different types of fun activities are presented in the ***Study Notebook***. These help reinforce the concepts students are learning and include making observations, some simple experiments, matching, fill in the blank, cut and paste, writing, following directions, and more.
- The one final and two midterm ***Quizzes*** are self-explanatory. For those who are not fans of quizzes, students can use the self-test at the end of the ***Lesson Plan*** instead.
- Another type of teaching aid is provided in the ***Graphics Package,*** which has two images from each chapter of the ***Student Textbook.***

Want to see more? An additional free downloadable chapter of these materials can be found on our website, where you can learn about our complete curriculum offerings. Schools and distributors: please contact us for information about our large order discounts and sample sets of entire books.

www.realscience4kids.com • office@gravitaspublications.com • (505)-266-2761

Focus On

Grades K-4

Elementary

3rd Edition

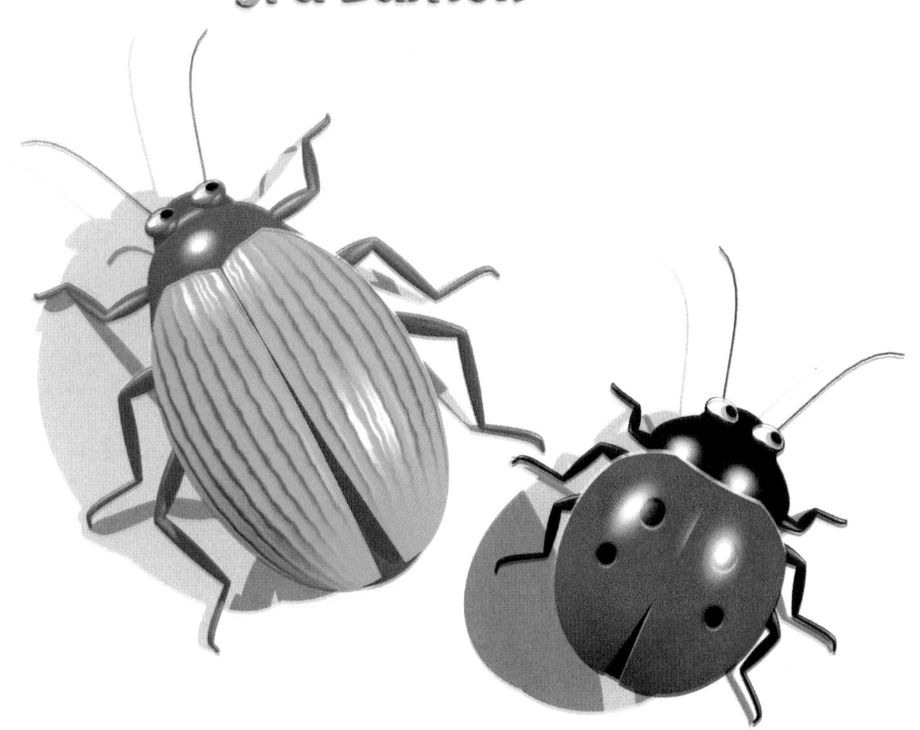

Rebecca W. Keller, PhD

Contents

CHAPTER 1 WHAT IS BIOLOGY? 1

1.1 Introduction 2
1.2 History of Biology 3
1.3 Modern Biology 4
1.4 Everyday Biology 6
1.5 Summary 8
1.6 Some Things to Think About 8

CHAPTER 2 BIOLOGIST'S TOOLBOX 9

2.1 Introduction 10
2.2 Brief History 10
2.3 Basic Biology Tools 12
2.4 Advanced Biology Tools 13
2.5 Summary 15
2.6 Some Things to Think About 15

CHAPTER 3 LIFE 17

3.1 Studying Life 18
3.2 Sorting Living Things 20
3.3 Domains and Kingdoms 23
3.4 Sorting Within Kingdoms 25
3.5 Naming 26
3.6 Summary 27
3.7 Some Things to Think About 28

CHAPTER 4 THE CELL: A TINY CITY 29

4.1 Creatures 30
4.2 The Cell 31
4.3 A Tiny City 33
4.4 Parts of a Cell 36
4.5 Summary 39
4.6 Some Things to Think About 39

CHAPTER 5 VIRUSES, BACTERIA, AND ARCHAEA 40

5.1 Introduction 41
5.2 Viruses 42
5.3 Bacteria 43
5.4 Archaea 47
5.5 Summary 48
5.6 Some Things to Think About 48

CHAPTER 6 PROTISTS MOVE 49

6.1 Tiny Creatures 50
6.2 Different Ways Protists Move 51
6.3 Summary 53
6.4 Some Things to Think About 54

CHAPTER 7 PROTISTS EAT 55

7.1 Euglena Eat 56
7.2 Paramecia Eat 57
7.3 Amoebas Eat 58
7.4 Other Protists Eat 59
7.5 Summary 60
7.6 Some Things to Think About 60

CHAPTER 8 FUNGI, MOLDS, MUSHROOMS, YEASTS 61

8.1 Introduction 62
8.2 Molds 63
8.3 Mushrooms 64
8.4 Yeasts 65
8.5 Summary 66
8.6 Some Things to Think About 66

CHAPTER 9 PLANTS 67

9.1 Introduction 68
9.2 So Many Plants! 68
9.3 Where Plants Live 69
9.4 Plant Cells 71
9.5 Summary 71
9.6 Some Things to Think About 72

CHAPTER 10 FOOD FOR PLANTS 73
10.1 Introduction 74
10.2 Factories 75
10.3 How Plants Make Food 76
10.4 Food Factories 76
10.5 Different Leaves 77
10.6 Summary 78
10.7 Some Things to Think About 78

CHAPTER 11 PLANT PARTS 80
11.1 Introduction 81
11.2 In the Ground: The Roots 82
11.3 Above the Soil: Leaves, Stems, Flowers 83
11.4 Other Places Plants Live 83
11.5 Summary 84
11.6 Some Things to Think About 85

CHAPTER 12 GROWING A PLANT 86
12.1 The Beginning: Seeds 87
12.2 The Middle: Baby Plants 88
12.3 The Finish: Flowers and Fruit 89
12.4 Starting Again: The Life Cycle 90
12.5 Summary 91
12.6 Some Things to Think About 91

CHAPTER 13 WHAT ARE ANIMALS? 93
13.1 Introduction 94
13.2 What Is an Animal? 95
13.3 Animal Cells 97
13.4 Animal Phyla 98
13.5 Summary 101
13.6 Some Things to Think About 101

CHAPTER 14 SQUISHY, SPINY, SLIMY ANIMALS 102

14.1 Introduction 103
14.2 Sponges 104
14.3 Jellyfish 106
14.4 Worms 107
14.5 Snails and Octopuses 109
14.6 Sea Stars, Sand Dollars, Sea Urchins 112
14.7 Summary 115
14.8 Some Things to Think About 115

CHAPTER 15 STINGING, CRAWLING, SQUIRMING ANIMALS 117

15.1 Introduction 118
15.2 Basic Body Plan 118
15.3 Insects 120
15.4 Spiders 126
15.5 Lobsters, Shrimp, and Crabs 127
15.6 Summary 128
15.7 Some Things to Think About 129

CHAPTER 16 SWIMMING, FLYING, SCALY, FURRY ANIMALS 130

16.1 Introduction 131
16.2 Fish 131
16.3 Frogs 134
16.4 Reptiles 138
16.5 Birds 141
16.6 Mammals 142
16.7 Summary 143
16.8 Some Things to Think About 144

GLOSSARY-INDEX 145

Chapter 7 Protists Eat

7.1 Euglena Eat

A euglena uses sunlight to make its own food. It changes sunlight to food by using chloroplasts. Chloroplasts are special parts inside a cell. A chloroplast contains chlorophyll which is a green substance that captures sunlight. Because chlorophyll is green, it gives euglena their green color.

A euglena will swim toward the sunlight that it uses to make food. A euglena has a little eyespot that helps it know where to find the sunlight.

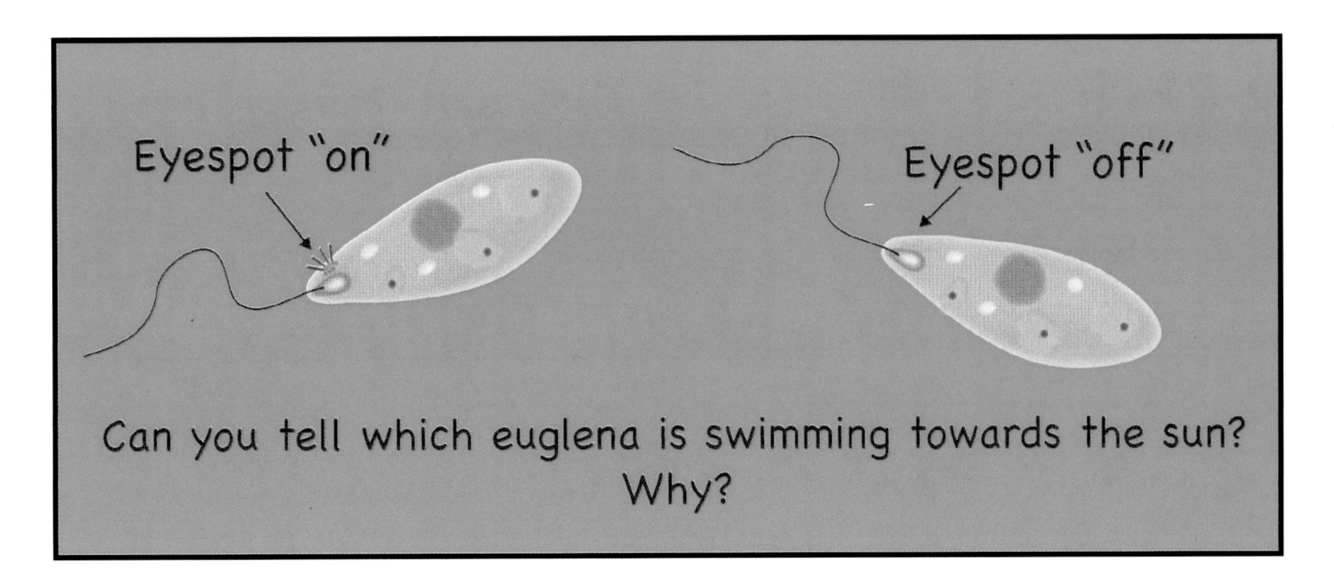

Can you tell which euglena is swimming towards the sun? Why?

7.2 Paramecia Eat

Not all protists can make their own food like a euglena does. A paramecium has to go find its food, just like we do! But a paramecium cannot go to the grocery store for eggs and milk like we can. It must swim around with its cilia to look for food in the water.

A paramecium eats other small creatures, such as other protozoa or bacteria. The paramecium has a mouth that it uses to capture food. The mouth does not move like a human mouth and it doesn't have any teeth.

The cilia around the mouth move, or beat, rapidly. This makes the water near the mouth of the paramecium swirl. Take your hands and move them in some water, and you can feel the water swirling around your hands.

A paramecium uses the swirling water to move food toward its mouth. When the food enters the mouth of the paramecium, it travels through a small tube into a tiny stomach and gets digested. The paramecium takes what it needs from the digested food, and the unused food is pushed out through a small hole.

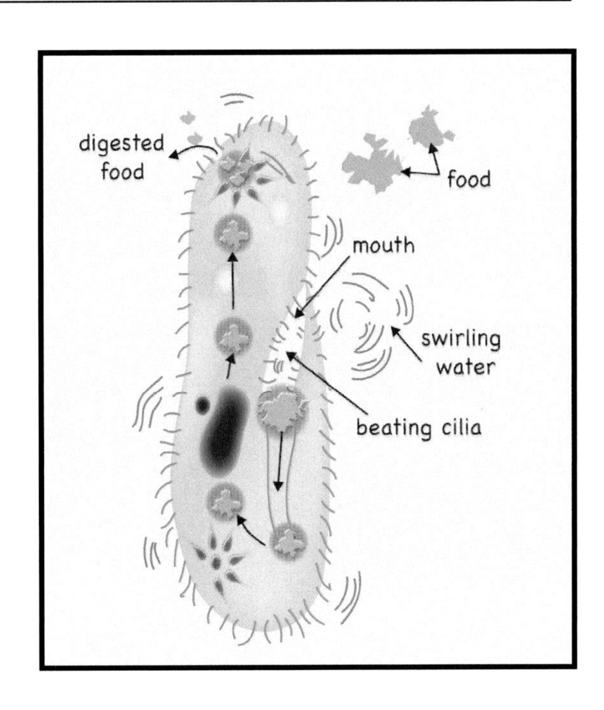

Although a paramecium has only one cell, it can move, eat, and digest food just like larger creatures. For being so small, the paramecium is an amazing creature.

7.3 Amoebas Eat

An amoeba eats with its feet! Can you imagine eating with your feet? It would be pretty hard for you to eat with your feet, but it isn't hard at all for an amoeba.

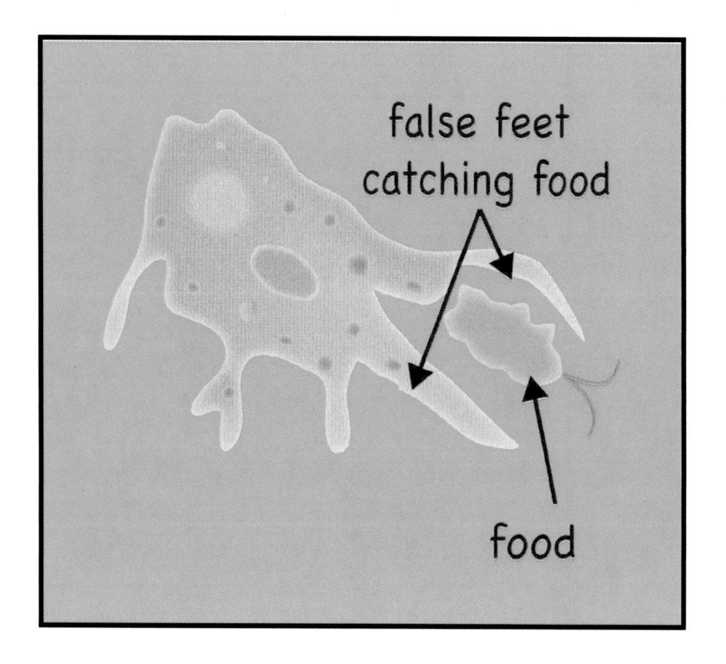

An amoeba uses its false feet, or pseudopods, to surround the food it wants to eat. Once the food is surrounded, the amoeba brings its feet together and makes something like a stomach out of the false feet that surround the food.

The stomach then absorbs the food into the body of the amoeba. That is how the amoeba eats with its feet!

7.4 Other Protists Eat

There are other protists that eat in other ways. For example, a protist called a *Coleps* rotates its whole body to swim through the water.

It also rotates as it eats. As it rotates, it uses its sharp teeth to bore through the food like a tiny drill. Then it eats the food it removes from the hole it has made with its teeth.

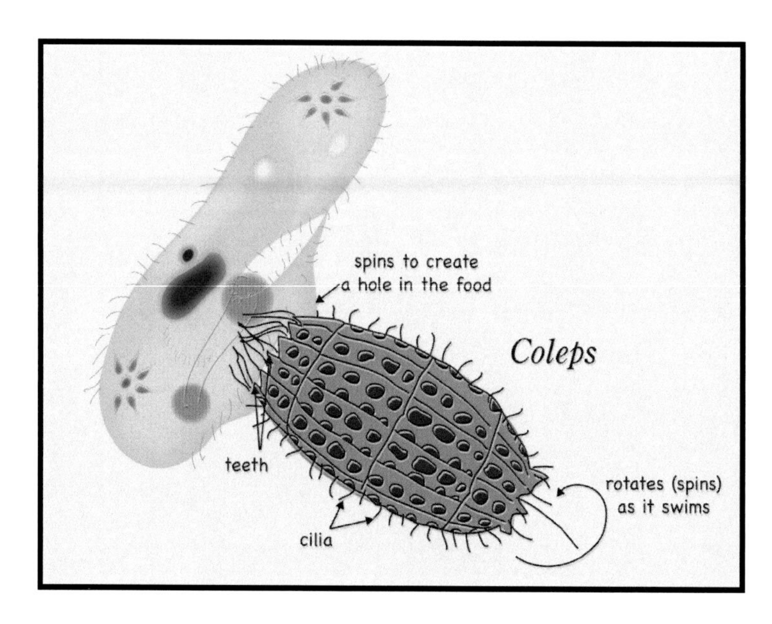

7.5 Summary

- A euglena uses chloroplasts to make its food.
- A paramecium uses cilia to swirl the water and sweep food into its mouth.
- An ameoba uses its pseudopods, or false feet, to capture food and eat.
- Other protists use other ways to eat.

7.6 Some Things to Think About

- What do you think would happen if a euglena did not have an eyespot?
- How would you describe the way a paramecium eats?
- Why do you think pseudopods are called false feet?
- How is the way a Coleps eats different from the way a paramecium eats? How is it different from the way an amoeba eats?

Grades K-4

Focus On Elementary

Laboratory Notebook

3rd Edition

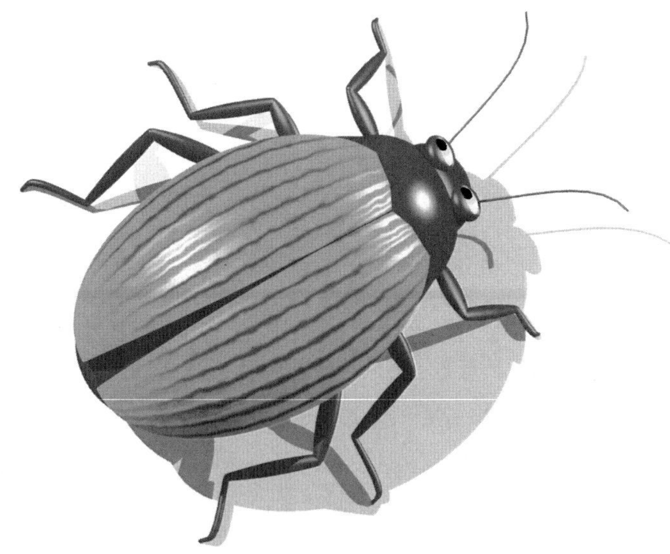

Rebecca W. Keller, PhD

A Note From the Author

Hi!

In this curriculum you are going to learn the first step of the scientific method:

Making good observations!

In the science of biology, making good observations is very important.

Each experiment in this notebook has several different sections. In the section called *Observe It*, you will be asked to make observations. In the *Think About It* section you will answer questions. There is a section called *What Did You Discover?* where you will write down or draw what you observed from the experiment. And finally, in the section *Why?* you will learn about the reasons why you may have observed certain things during your experiment.

These experiments will help you learn the first step of the scientific method and..... they're lots of fun!

Enjoy!

Rebecca W. Keller, PhD

Contents

Experiment 1: WHAT IS LIFE? 1

Experiment 2: TAKING NOTES 9

Experiment 3: WHERE DOES IT GO? 23

Experiment 4: WHAT DO YOU NEED? 38

Experiment 5: YUMMY YOGURT 48

Experiment 6: LITTLE CREATURES MOVE 53

Experiment 7: LITTLE CREATURES EAT 65

Experiment 8: OLDY MOLDY 77

Experiment 9: NATURE WALK: OBSERVING PLANTS 85

Experiment 10: WHO NEEDS LIGHT? 92

Experiment 11: THIRSTY FLOWERS 105

Experiment 12: GROWING SEEDS 116

Experiment 13: NATURE WALK: OBSERVING ANIMALS 129

Experiment 14: RED LIGHT, GREEN LIGHT 136

Experiment 15: BUTTERFLIES FLUTTER BY 148

Experiment 16: TADPOLES TO FROGS 160

Experiment 7

Little Creatures Eat

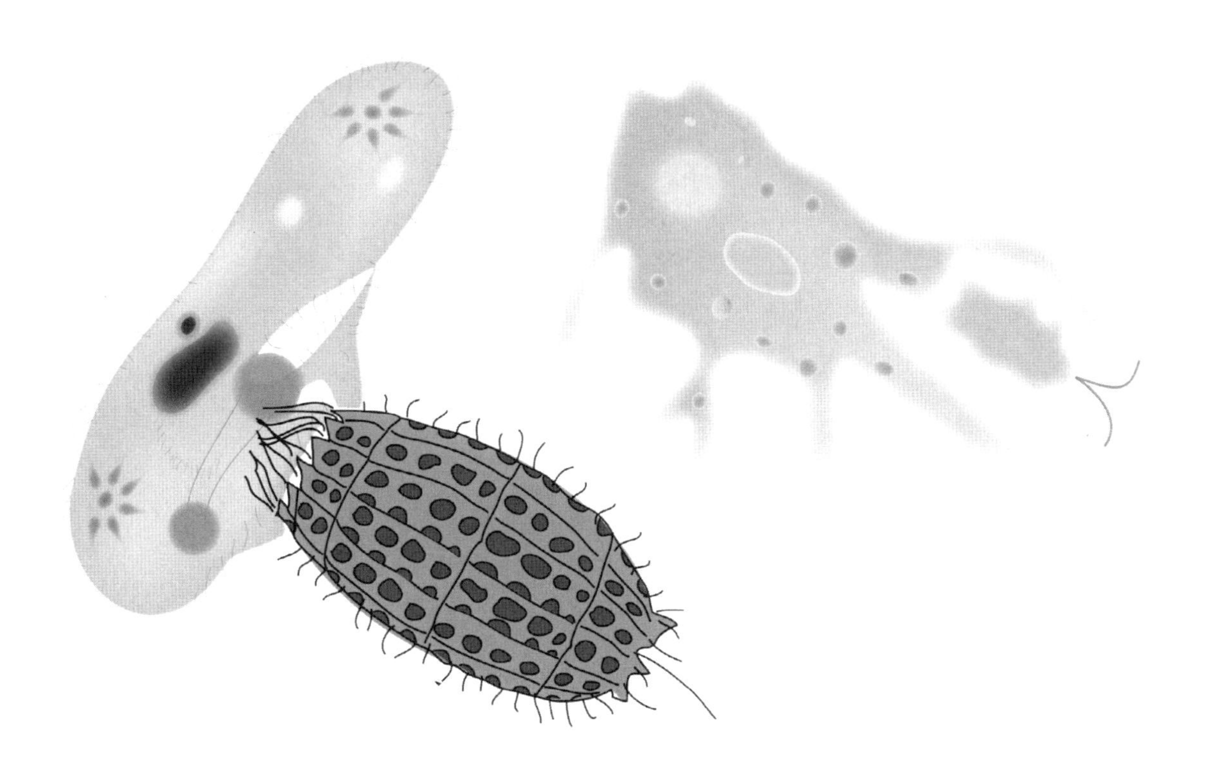

Introduction

In this experiment you will take another look at very tiny creatures and observe how they eat.

I. Think About It

If you use a microscope to look at a protist eating, what do you think you will see? Draw what you think you will see.

II. Observe It

1. Take some pond water and put it under the microscope. Observe whether any protists are eating. Draw what you see.

❷ See if you can observe different creatures that are eating. Draw one below.

❸ Draw a different creature that is eating.

❹ Draw another different creature that is eating.

❺ Draw the food one creature might be eating.

❻ Are there two creatures that are eating in the same way? Draw them below.

❼ Are there two creatures that are eating in different ways? Draw them below.

III. What Did You Discover?

❶ Did the protists eat like you thought they would? Why or why not?

---

❷ What was the first thing you noticed about the eating protists?

---

❸ Was there anything you did not expect to find while you were watching the protists eat? Describe it.

---

❹ How many different ways did they eat?

---

❺ Describe your favorite creature. Explain why it is your favorite.

IV. Why?

Protists eat in different ways. Some protists make their own food, like the green euglena does. Some protists use their tiny hairs to sweep food into their mouths, like the paramecium does. And other protists capture their food with their feet, like the amoeba does. Because there are lots of different kinds of protists, there are lots of different ways protists eat.

Protists can eat lots of different kinds of food. They can eat algae or other small plants. They can eat yeast, and they can eat other protists. Imagine what might happen when two protist-eating protists meet. Who gets to eat whom?

You also eat, but like most humans, you usually use your mouth to eat. Humans can eat both plants and animals. You usually don't need to hunt for your food—unless your brother steals your piece of cake! But you do need food. You are not like a euglena who can make its own food, and you can't catch your food with your feet like an amoeba. In a city, you need to rely on other people who can provide you with the food you need. You need milk from the dairy and bread from the bakery and eggs from the farm and just for fun—chocolate from the chocolate factory! A protist doesn't have other protists finding food for it—it must find its own food.

V. Just For Fun

Do protists like chocolate? Try it. Place a small piece of chocolate on the slide and see if protists eat it. Record your observations.

Protists and Chocolate

Grades K-4

Focus On Elementary

Teacher's Manual

3rd Edition

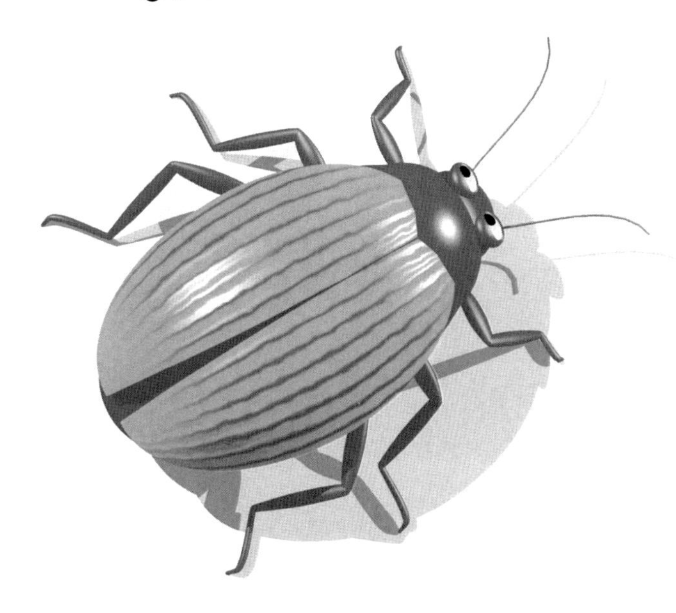

Rebecca W. Keller, PhD

A Note From the Author

This curriculum is designed to provide an introduction to biology for students in the elementary level grades. *Focus On Elementary Biology—3rd Edition* is intended to be used as the first step in developing a framework for the study of real scientific concepts and terminology in biology. This *Teacher's Manual* will help you guide students through the series of experiments in the *Laboratory Notebook*. These experiments will help the students develop the skills needed for the first step in the scientific method—making good observations.

There are several sections in each chapter. The section called *Observe It* helps the students explore how to make good observations. The *Think About It* section provides questions for the students to think about and use to make further observations. In every chapter there is a *What Did You Discover?* section that gives the students an opportunity to summarize the observations they have made. A section called *Why?* provides a short explanation of what students may or may not have observed. And finally, in each chapter there is a section called *Just For Fun* that contains an additional activity.

The experiments take up to 1 hour. The materials needed for each experiment are listed on the next page and also at the beginning of each experiment.

Enjoy!

Rebecca W. Keller, PhD

Materials at a Glance

Experiment 1	Experiment 3	Experiment 4	Experiment 5	Experiment 6
non-living object to observe (such as a rock or piece of wood) living thing to observe (such as an ant, frog, bird, cat, or dog) colored pencils **Experiment 2** magnifying glass colored pencils	cotton balls rubber ball tennis ball banana apple rocks Legos other objects colored pencils	internet access and/or reference books colored pencils	milk, .25 l (1 cup) plain yogurt, .5 liter (2 cups) fork spoon cups or small bowls (several) food items such as honey, berries, chopped fruit or vegetables, spices, herbs, cocoa, chocolate chips, etc. (*Just For Fun* section)	microscope with a 10x or 20x objective lens (see the following How to Buy a Microscope section) plastic microscope slides[1] eye dropper pond water or protozoa kit[1] Protists (protozoa) can also be observed in hay water. To make hay water, cover a clump of dry hay with water and let it stand for several days at room temperature. Add water as needed

Experiment 7	Experiment 8	Experiment 9	Experiment 10	Experiment 11
(see Experiment 6) small piece of chocolate **Optional** baker's yeast Eosin Y stain[2] distilled water	6-8 sealable plastic bags waterproof disposable gloves piece of newspaper or plastic 2 pieces of fruit 2-3 pieces of bread (works best if bread does not have preservatives) marking pen water **Optional** colored pencils	notebook or drawing pad with blank pages (not ruled) to make a nature journal pencil colored pencils **Optional** camera and printer tape backpack snack and bottle of water	2 small houseplants of the same kind and size 2 more small houseplants of the same kind and size water measuring cup closet or cardboard box colored pencils	2-4 white carnations 1 or more other white flowers (rose, lily, etc.) 2-3 small jars food coloring water tape knife colored pencils **Optional** magnifying glass

[1] As of this writing, the following materials are available from Home Science Tools, www.hometrainingtools.com: plastic microscope slides, MS-SLIDSPL or MS-SLPL144, Basic Protozoa Set, LD-PROBASC

[2] Eosin Y stain, CH-EOSIN (Home Science Tools)

Experiment 12	Experiment 13	Experiment 14	Experiment 15	Experiment 16
1-2 small clear glass jars 2 or more dried beans (white, pinto, soldier, etc.) 2 or more additional dried beans (different kind) or other seeds absorbent white paper scissors knife plastic wrap clear tape rubber band water **Optional** magnifying glass	student's field notebook pencil, pen colored pencils **Optional** camera and printer tape backpack snack and bottle of water	large tray or plastic box, at least .3 m (1 ft.) on each side, and cover garden dirt (with lots of organic material) spoon or garden trowel 12 snails/slugs and/or 20–40 worms [3] holding box for the snails/worms to keep them moist and dark water experimental snail and worm barriers. Set the amount you are going to use in an open container in the sun for a few days. table or rock salt plus three of the following: cinnamon baking soda black pepper cornstarch flour borax an active anthill	butterfly kit small cage Butterfly kits can be purchased from a variety of different sources, such as: Home Science Tools: www. hometrainingtools. com Insect Lore: www.insectlore. com	tadpole kit (or tadpoles or frog eggs collected locally) A tadpole kit can be purchased from Home Science Tools: www. hometrainingtools. com. aquarium water tadpole food

[3] Look for online or local sources of snails and/or earthworms. Or you and your students may be able to collect them yourselves.

Materials: Quantities Needed for All Experiments

Equipment

aquarium
cage, small
cup, measuring
cups or small bowls (several)
eye dropper
fork
jars, 2-3 small, clear glass
knife
Legos
magnifying glass
microscope with a 10x or 20x objective lens[1]
scissors
spoon
spoon or garden trowel
tray or plastic box, large, at least .3 m (1 ft.) on each side, and cover

Optional

camera and printer
magnifying glass
camera
backpack

Foods

apple
banana
beans, dried (white, pinto, soldier, etc.), 2 or more
beans, dried (different from above) or other seeds, 2 or more
bread, 2-3 pieces (best without preservatives)
chocolate, small piece
food items such as honey, berries, chopped fruit or vegetables, spices, herbs, cocoa, chocolate chips, etc.
fruit, 2 pieces
milk, .25 l (1 cup)
yogurt, plain, .5 liter (2 cups)

Optional

baker's yeast
snack and bottle of water

Materials

ball, rubber
ball, tennis
box for snails/worms to keep them moist and dark
butterfly kit[2]
carnations, 2-4 white
cotton balls
dirt, garden (with lots of organic material)
flowers (rose, lily, etc.), white, 1 or more (not carnations)
food coloring
gloves, waterproof, disposable
houseplants, 2 small - same kind and size
houseplants, 2 additional, small - same kind and size
living thing to observe (such as an ant, frog, bird, cat, or dog)
microscope slides, plastic[2]
newspaper or plastic, 1 piece
notebook or drawing pad with blank pages (not ruled)
non-living object to observe (such as a rock or piece of wood)
objects, misc.
paper, absorbent white
pen
pen, marking
pencil
pencils, colored
plastic bags, sealable, 6-8
plastic wrap
pond water or protozoa kit
protists (protozoa)[2]
rocks
rubber band
snail and worm barriers, student choice of materials
snails/slugs, 12, and/or 20–40 worms[3]

Materials (continued)

table or rock salt plus three of the following: cinnamon, baking soda, black pepper, cornstarch, flour, borax
tadpole food
tadpole kit (or tadpoles or frog eggs collected locally)[2]
tape
tape, clear
water

Optional

Eosin Y stain[2]
water, distilled

Other

anthill, active
closet or cardboard box
internet access and/or reference books

[1] See the following *How to Buy a Microscope* section for recommendations.
[2] As of this writing, the following materials are available from Home Science Tools, www.hometrainingtools.com:
Butterfly kit (can also be purchased from Insect Lore: www.insectlore.com)
Eosin Y stain, CH-EOSIN
Plastic microscope slides, MS-SLIDSPL or MS-SLPL144
Basic Protozoa Set, LD-PROBASC
Tadpole kit
[3] Look for online or local sources of snails and/or earthworms. Or you and your students may be able to collect them.

How to Buy a Microscope

What to Look For

- A metal mechanical stage.
- A metal body painted with a resistant finish.
- DIN Achromatic Glass objective lenses at 4X, 10X, 40X (a 100X lens is optional but recommended).
- A focusable condenser (lens that focuses the light on the sample).
- Metal gears and screws with ball bearings for movable parts.
- Monocular (single tube) "wide field" ocular lens.
- Fluorescent lighting with an iris diaphragm.

Price Range

$50-$150: Not recommended: These microscopes do not have the best construction or parts and are often made of plastic. These microscopes will cause frustration, discouraging students.

$150-$350: A good quality standard student microscope can be found in this price range. We recommend Great Scopes for a solid student microscope with the best parts and optics in this price range. http://www.greatscopes.com

Above $350: There are many higher end microscopes that can be purchased, but for most students these are too much microscope for their needs. However, if you have a child who is really interested in microscopy, wants to enter the medical or scientific profession, or may become a serious hobbyist, a higher end microscope would be a valuable asset.

Objective lenses: Magnification/Resolution/Field of View/Focal Length

The objective lenses are the most important parts of the microscope. An objective lens not only magnifies the sample, but also determines the resolution. However, higher powered objective lenses with better resolution have a smaller field of view and a shorter focal length.

The resolution and working distance (focal length) of a lens is determined by its numerical aperture (NA). Following is a list of magnifications, numerical aperture, and working distance for some common achromatic objective lenses.

How to Buy a Microscope (Continued)

Magnification	Numerical Aperture	Working Distance (mm)
4X	0.10	30.00
10X	0.25	6.10
20X	0.40	2.10
40X	0.65	0.65
60X	0.80	0.30
100X (oil)	1.25	0.18

You can see as the magnification increases the numerical aperture increases (which means the resolution increases) and the working distance decreases.

Choosing the right lens for the right sample is part of the art of microscopy.

Most student projects can be achieved with a 40X objective, however a 100X objective lens can be added to make observing bacteria and small cell structures possible.

Below is a general chart showing the recommended objective lens to use for different types of samples.

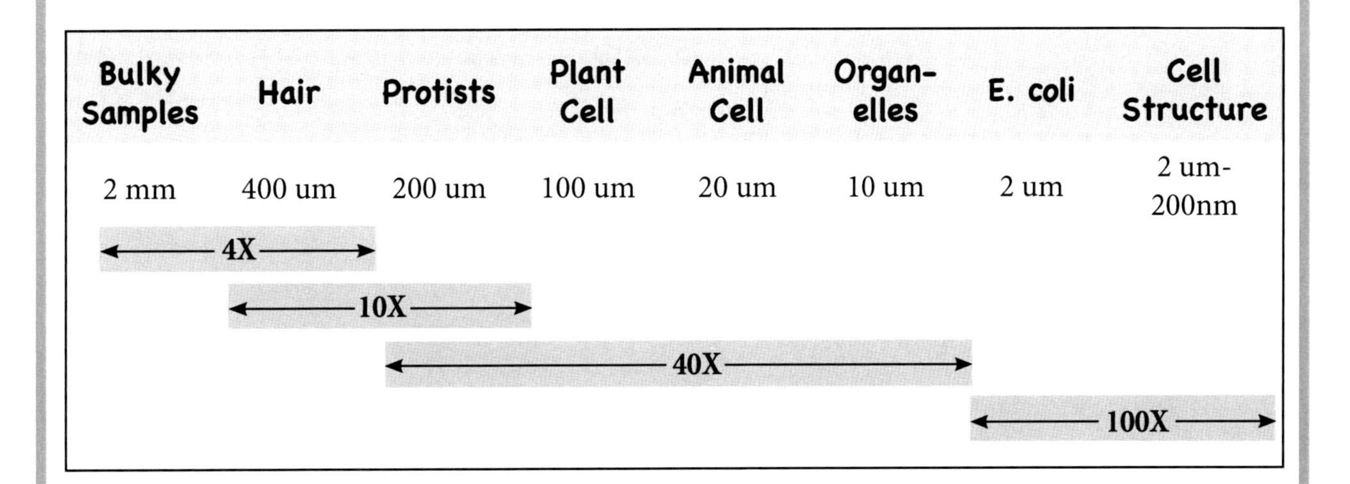

Contents

Experiment 1:	WHAT IS LIFE?	1
Experiment 2:	TAKING NOTES	4
Experiment 3:	WHERE DOES IT GO?	7
Experiment 4:	WHAT DO YOU NEED?	13
Experiment 5:	YUMMY YOGURT	18
Experiment 6:	LITTLE CREATURES MOVE	21
Experiment 7:	LITTLE CREATURES EAT	25
Experiment 8:	OLDY MOLDY	29
Experiment 9:	NATURE WALK: OBSERVING PLANTS	33
Experiment 10:	WHO NEEDS LIGHT?	36
Experiment 11:	THIRSTY FLOWERS	40
Experiment 12:	GROWING SEEDS	44
Experiment 13:	NATURE WALK: OBSERVING ANIMALS	48
Experiment 14:	RED LIGHT, GREEN LIGHT	51
Experiment 15:	BUTTERFLIES FLUTTER BY	55
Experiment 16:	TADPOLES TO FROGS	58

Experiment 7

Little Creatures Eat

Materials Needed

- microscope with a 10X or 20x objective lens (look online for sources such as Great Scopes or Carolina Biological Supply)
- plastic microscope slides
- eye dropper
- pond water or protozoa kit
- small piece of chocolate

Optional

- baker's yeast
- Eosin Y stain
- distilled water

Protists (protozoa) can also be observed in hay water. To make hay water, cover a clump of dry hay with water, and let it stand for several days at room temperature. Add water as needed.

As of this writing, the following materials are available from Home Science Tools, www.hometrainingtools.com:

- plastic microscope slides, MS-SLIDSPL or MS-SLPL144
- Basic Protozoa Set, LD-PROBASC
- Eosin Y stain, CH-EOSIN

Objectives

In this unit students will look at pond water, hay water, or water from a protozoa kit to observe how protists (protozoa) eat.

The objectives of this lesson are for students to:

- Make careful observations of protists eating.
- Practice using a microscope.

Experiment

In this experiment students will focus on protists that are eating. If pond water or hay water is being used, there should be plenty of food for the protists to eat.

Baker's yeast stained with Eosin Y can be added to any of the kinds of protozoa water. The Eosin Y stained yeast will be ingested by the protists. Once ingested, the red stained yeast will turn blue. It may take some time for this observation.

To make baker's yeast and Eosin Y stain:

- Add 5 milliliters (one teaspoon) of dried yeast to 120 milliliters (1/2 cup) of distilled water. Allow it to dissolve. Let the mixture sit for a few minutes, then add one dropper of Eosin Y to one dropper of the baker's yeast solution and let sit for a few minutes.
- Look at a droplet of the mixture under the microscope. You should be able to see individual yeast cells that are stained red.

I. Think About It

Read this section of the *Laboratory Notebook* with your students.

The students have read about how protists eat. Have them first think about what it might look like for a protist to eat. Help them explore their ideas with questions such as:

- *How do you think a paramecium eats?*
- *Do you think you can watch it eat?*
- *Do you think you can tell if the food is going inside?*
- *How do you think an amoeba eats?*
- *Do you think an amoeba can eat fast moving food? Why or why not?*
- *What else do you think you might see as the protists eat?*

Have the students draw what they think they will observe through the microscope as they watch protists eat.

II. Observe It

Read this section of the *Laboratory Notebook* with your students.

❶ a) Help the students place a small droplet of the protozoa solution onto a microscope slide.

b) If using Eosin Y stained baker's yeast, have the students add a droplet of the stained baker's yeast to the protozoa water on the slide.

c) Help the students carefully place the slide in the microscope.

d) Have the students look through the eyepiece at the protozoa water on the slide.

(You may also position the slide correctly in the microscope and then add the liquids to it.)

It is important for students to practice observing as many different details as possible. Have them draw what they observe.

❷-❺ There are several drawing frames in the *Laboratory Notebook* for students to fill in with drawings of their observations of protists eating. Encourage the students to spend plenty of time looking at all the different features they observe. You can encourage them to stay at the microscope by engaging them with questions such as:

- *What kind of protist do you think you are seeing?*
- *Is it eating?*
- *Can you tell what it is eating?*
- *Can you tell if the protist is eating another protist or something else?*
- *How fast does it eat?*

❻-❼ Have the students compare some of the protists they are observing. They can make comparisons between different protists of the same kind (two paramecia, for example) and protists of different kinds (possibly a paramecium and an amoeba).

III. What Did You Discover?

Read this section of the *Laboratory Notebook* with your students.

Have students answer the questions about the protists they observed. Encourage them to summarize their answers based on their observations. They should have been able to see different protists eating. Have them explain what their favorite protist was, how it was eating, and why it was their favorite. Help them notice any differences between what they thought they would observe and what they actually observed.

IV. Why?

Read this section of the *Laboratory Notebook* with your students.

Different protists eat in different ways. Your students may or may not have been able to observe the protists eating. Explain to them that watching protists eat is sometimes like watching the tigers eat at the zoo. They may not be hungry. Repeat the experiment at a different time if your student was not able to observe protists eating.

V. Just For Fun

Help the students put a tiny piece of chocolate on the slide with the protozoa water. Have them look through the microscope to see if the protozoa will eat chocolate. Have them record their observations in the space provided.

Grades K-4

LESSON PLAN

3rd Edition

Rebecca W. Keller, PhD

LESSON PLAN INSTRUCTIONS

This *Lesson Plan* accompanies the *Focus On Elementary Biology Student Textbook, Laboratory Notebook,* and *Teacher's Manual— 3rd Edition*. It is designed to be flexible to accommodate a varying schedule as you go through the year's study. And it makes it easy to chart weekly study sessions and create a portfolio of your student's yearlong performance. The PDF format allows you to print pages as you need them.

This Lesson Plan file includes:

- Weekly Sheets
- Sticker Templates
- Self-Review Sheet
- Self-Test Sheet

Materials recommended but not included:

- 3-ring binder
- Indexing dividers (3)
- Labels—24 per sheet,1.5" x 1.5" (Avery 22805)

Use the Weekly Sheets to map out daily activities and keep track of student progress. For each week you decide when to read the text, do the experiment, explore the optional connections, review the text, and administer tests. For those families and schools needing to provide records of student performance and show compliance to standards, there is a section on the Weekly Sheets that shows how the content aligns to the National Science Standards.

To use this Lesson Plan:

- Print the Weekly Sheets
- Print Self-Review Sheets
- Print Self-Test Sheets
- Print the stickers on 1.5" x 1.5" labels
- Place all the printed sheets in a three-ring binder separated by index dividers

At the beginning of each week, use the squares under each weekday to plan your daily activities. You can attach printed stickers to the appropriate boxes or write in the daily activities. At the end of the week, use the Notes section to record student progress and performance for that week.

WEEKLY LESSON PLAN SAMPLES

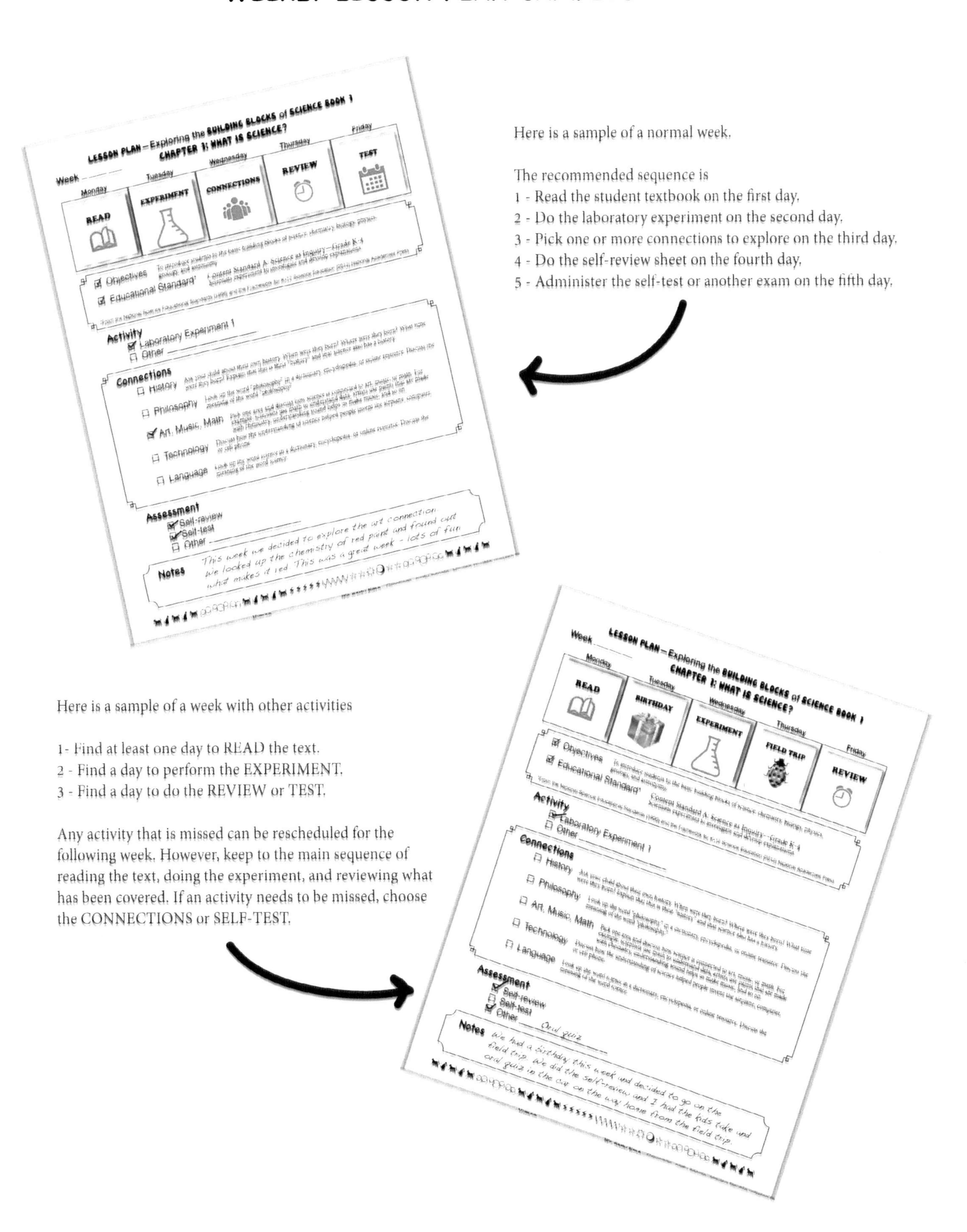

Here is a sample of a normal week.

The recommended sequence is

1 - Read the student textbook on the first day.
2 - Do the laboratory experiment on the second day.
3 - Pick one or more connections to explore on the third day.
4 - Do the self-review sheet on the fourth day.
5 - Administer the self-test or another exam on the fifth day.

Here is a sample of a week with other activities

1- Find at least one day to READ the text.
2 - Find a day to perform the EXPERIMENT.
3 - Find a day to do the REVIEW or TEST.

Any activity that is missed can be rescheduled for the following week. However, keep to the main sequence of reading the text, doing the experiment, and reviewing what has been covered. If an activity needs to be missed, choose the CONNECTIONS or SELF-TEST.

Week ________

CHAPTER 7: PROTISTS EAT

Monday	Tuesday	Wednesday	Thursday	Friday

☑ Objectives To introduce students to the microscopic organisms called protists.

☑ Educational Standard* Content Standard C: Life Science: Grade K-4
Organisms have basic needs.

*From the National Science Educational Standards (1996) and the Framework for K-12 Science Education (2012) National Academies Press

Activity

☐ Laboratory Experiment 7
☐ Other ________________

Connections

☐ History Continue to explore the history of protists. What were the first protists to be discovered?

☐ Philosophy Look up Anton van Leeuwenhoek and continue to explore how his discovery changed the way we understand the world around us.

☐ Art, Music, Math Explore the intricate designs of protists and how these designs have inspired artists and architects.

☐ Technology What technologies have allowed us to learn more about protists? Discuss how learning about protists can be useful to us.

☐ Language Look up the word *pseudopod* in a dictionary or encyclopedia. Discuss the meaning of the word *pseudopod*.

Assessment

☐ Self-review
☐ Self-test
☐ Other ________________

Notes

SELF-REVIEW

Think about all of the ideas, concepts, and facts you read about in this chapter. In the space below, write down everything you've learned.

Date ____________________ Chapter ____

SELF-TEST

Imagine you are the teacher and you are giving your students an exam. In the space below, write 5 questions you would ask a student based on the information you learned in this chapter.

Date ________________ Chapter ____

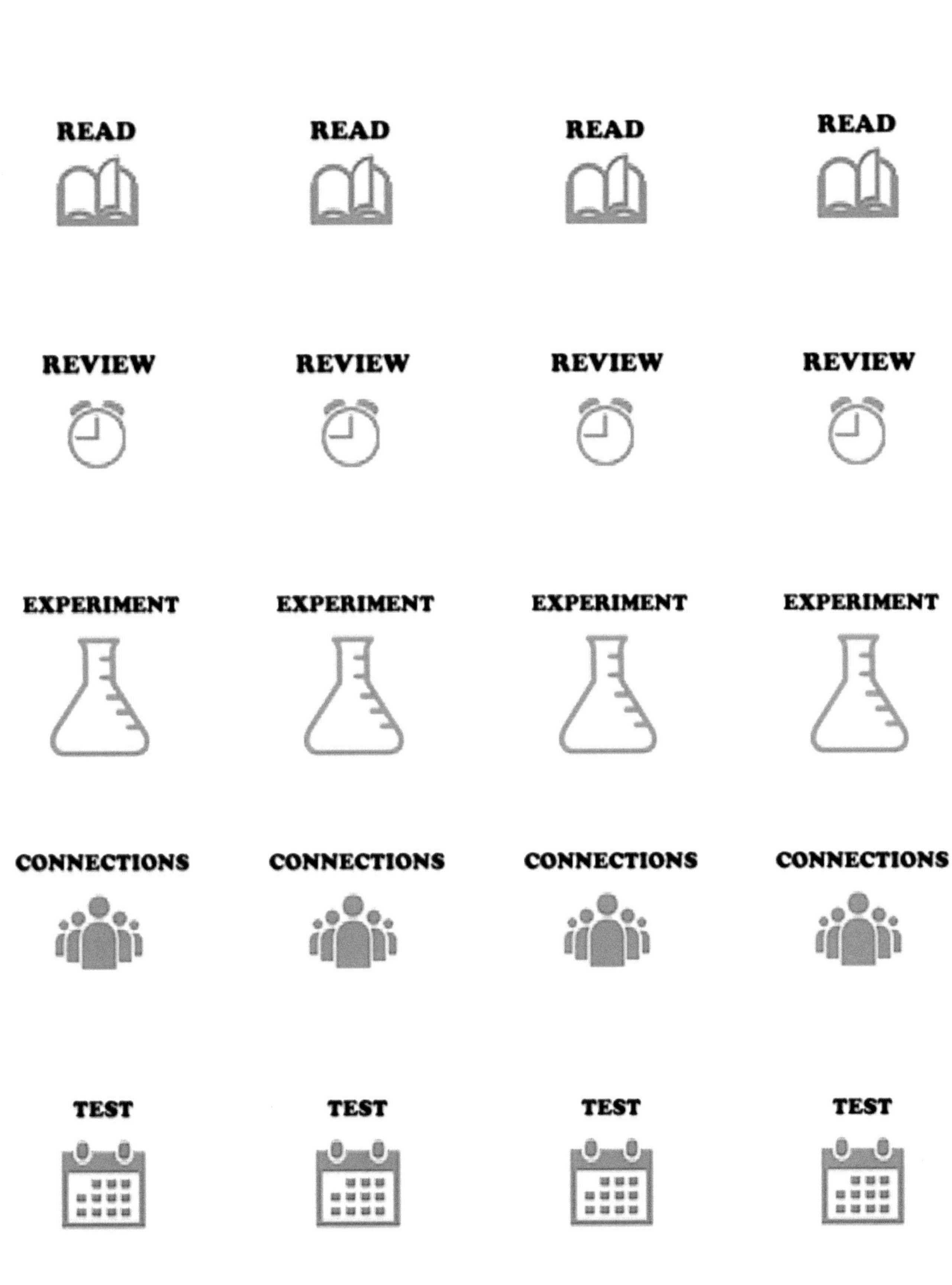
READ
READ
READ
READ
REVIEW
REVIEW
REVIEW
REVIEW
EXPERIMENT
EXPERIMENT
EXPERIMENT
EXPERIMENT
CONNECTIONS
CONNECTIONS
CONNECTIONS
CONNECTIONS
TEST
TEST
TEST
TEST

READ

READ

READ

READ

HOLIDAY	HOLIDAY	HOLIDAY	HOLIDAY

FIELD TRIP	FIELD TRIP	FIELD TRIP	FIELD TRIP

BIRTHDAY	BIRTHDAY	BIRTHDAY	BIRTHDAY

REST DAY	REST DAY	REST DAY	REST DAY

REST DAY	REST DAY	REST DAY	REST DAY

SICK DAY	SICK DAY	SICK DAY	SICK DAY
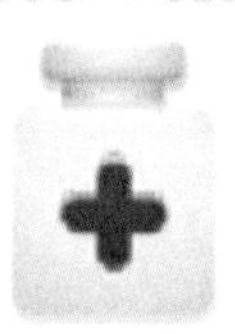			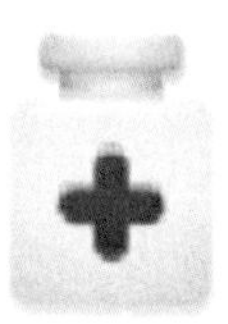

Grades K-4

ELEMENTARY

STUDY NOTEBOOK

3rd Edition

Rebecca W. Keller, PhD

FOCUS ON ELEMENTARY BIOLOGY STUDY NOTEBOOK

This Study Notebook has activities for you to do that will help you learn the ideas presented in each chapter of the Student Textbook.

Materials you will need

- 8.5 x 11 white paper
- color printer
- scissors
- glue or clear tape
- colored pencils
- 1 manila file folder
- 3 brad paper fasteners or 3-ring binder
- 3-hole punch

STEP 1 Printing

- Download the Study Notebook file for the chapter you are reading.
- Use the printer settings: portrait, letter, 8.5 x 11.
- Print the pages single sided.

STEP 2 Activities

- The little blue boxes on the left-hand side of the main pages show you which section of the Student Textbook has the information for that activity.
- For the chapter you are studying, do the activities on the two main pages (those that have page numbers at the bottom): fill in the blanks, answer questions, and follow the directions for other types of activities.
- On the **Stuff to Cut Out** pages, follow the directions for cutting out the pieces and gluing or taping them to the main pages.

STEP 3 Make the Study Notebook pages into a Book

- Cut the file folder in half along the fold.
- Use a 3-hole punch to make holes along the cut edge of the file folder pieces.

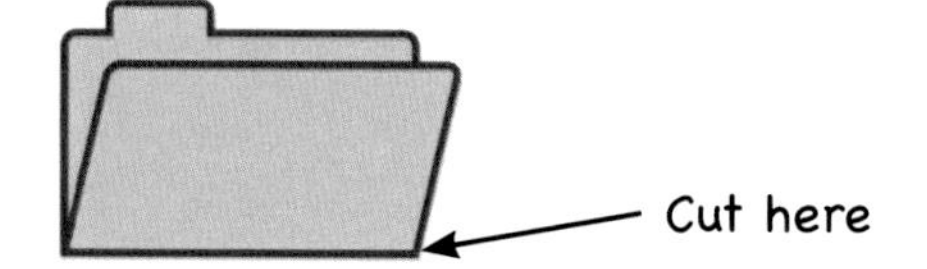

- Use the two pieces for the front and back covers.
- As you complete each chapter, punch holes in the pages and insert them between the front and back covers of your Study Notebook.

This is YOUR book! Add color to the pages along with doodles, squiggles, and notes in the margins. The backs of the pages are great for writing observations and ideas. Add your own pages with more ideas, observations, questions, science news you have heard about, and anything else you want to remember.

7. Protists Eat!

7.1

What do **PROTISTS** eat and **how** do they eat it?

It is silly to think that **any** **PR _ T _ _T** can make its own **f _ _ d**! YES! NO!

7.1

EUGLENA FACTS!

(Put a **T** in front of statements that are **true** and an **F** in front of statements that are **false**.)

____ Euglena use sugars to make their own food.

____ Euglena use sunlight to make their own food.

____ Euglena don't make their own food.

____ Chlorophyll is a red substance.

____ Chlorophyll gives a euglena its color.

____ An eyelid helps the euglena find sunlight.

7.2

THE AMAZING PARAMECIUM!

How is a **paramecium's mouth** different from **YOUR** mouth?

How does a paramecium get its food?
(Circle your answers.)

Swirls water — Uses a flagellum — Shops

Uses false feet — Uses cilia — Swims around

Uses a mouth — Uses chlorophyll — Uses a fork

What are some things that make a paramecium an amazing creature?

Punch holes on this edge.

Punch holes on this edge.

7.3

Do YOU think it is **Disgusting** to an **amoeba** that it has to eat with its **feet?**

NO! Because it uses its

p _ _ _ _ _ _ _ _ _ _ which are

F _ _ _ _ **F** _ _ _ !

(Draw an amoeba eating.)

7.4

How is a **COLEPS** that is **SWIMMING** like a **COLEPS** that is **EATING?**

(Draw a *Coleps*.)

7.1
7.2
7.3

(See directions In ***Stuff to Cut Out*** section.)

Glue **TAB 7A** Here

A DAY IN THE LIFE OF AN AMOEBA

(Write a short story about a day in the life of an amoeba.)

Stuff to Cut Out for Chapter 7

Make a little book!

Cut out the book pages found below and on the next page, and then write stories about protists. You can write on both sides of the little pages and you can cut out blank pages to add if you need more space. For the amoeba story on page 14, you can continue the story on the back of that page.

To begin your story for a particular protist, think about what you have learned about the protist. Then ask yourself questions about what you think life might be like for that protist. For example: How does it swim? Where does it go? What does it see? What does it eat and when? Are there other protists around? What does it do at night? Think of more questions and use your imagination in writing your stories!

When you have finished the stories, put the pages in order and then glue them, last page first, onto page 14. If you'd like, you can have a friend or your teacher read your stories and then take a quiz you have written. The completed quiz can be fastened into your Study Notebook.

Cut out the **Book Page** below on its solid outline and write your story.
Then match **yellow TAB 7A** to the **green Glue TAB 7A Here** on page 14.

Glue **TAB 7B** Here

TAB 7A

A DAY IN THE LIFE OF A EUGLENA

(Write a short story about a day in the life of a euglena.)

____

____

____

____

____

____

____

____

More Stuff to Cut Out for Chapter 7

Cut out this **Book Page** on its solid outline and write your story. Then match **yellow TAB 7B** to the **green Glue TAB 7B Here** on page 14.

Glue **TAB 7C** Here

TAB 7B

A DAY IN THE LIFE OF A PARAMECIUM

(Write a short story about a day in the life of a paramecium.)

Make a cover for your stories about protists. In the box below, fill in a title for your little book and draw the three main characters of your stories. Then cut out this piece on its solid outline and match **yellow TAB 7C** to the **green Glue TAB 7C Here** on page 14.

TAB 7C

Name ____________________________ **Date** __________________

Focus On Elementary Biology 3rd Edition, Midterm 1
Chapters 1-8, 20 questions, 10 points each

15. How does a paramecium get food? (10 points)
    - ◯ It catches light from the Sun and uses it to make its own food.
    - ◯ It uses its eyespot to find food and its flagellum to swim toward food.
    - ◯ It swims with cilia to find food and swirl the food into its mouth.
    - ◯ It uses false feet to surround the food.

16. Because a euglena needs sunlight to make its own food, it uses a sunspot to locate the light. (10 points)
    - ◯ True
    - ◯ False

17. A chloroplast... (Check all that apply.) (10 points)
    - ☐ Is the same as an eyespot.
    - ☐ Contains chlorophyll.
    - ☐ Is used to make food.
    - ☐ Makes amoebas green in color.
    - ☐ Captures sunlight.
    - ☐ Is found in euglena.
    - ☐ Is found in paramecia.

Focus On Elementary Biology 3rd Edition, Final Quiz
Chapters 1-16, 32 questions, 10 points each

13. A euglena eats by using cilia to swirl food into its mouth. (10 points)
    - ◯ True
    - ◯ False

14. Check all the statements that are true for protists. (10 points)
    - ☐ All protists can make their own food.
    - ☐ A paramecium eats other small creatures.
    - ☐ Protists use cilia, flagella, or pseudopods to help them get food.
    - ☐ Coleps rotates its body to swim and to eat.
    - ☐ A euglena eats paramecia.
    - ☐ A euglena uses chlorophyll to make its own food.
    - ☐ An amoeba surrounds its food with pseudopods.

Focus On Elementary Biology 3rd Edition, Midterm 1
Chapters 1-8, 20 questions, 10 points each

15. It swims with cilia to find food and swirl the food into its mouth.
16. False
17. Contains chlorophyll., Is used to make food., Captures sunlight., Is found in euglena.

Focus On Elementary Biology 3rd Edition, Final Quiz
Chapters 1-16, 32 questions, 10 points each

13. False
14. A paramecium eats other small creatures., Protists use cilia, flagella, or pseudopods to help them get food., Coleps rotates its body to swim and to eat., A euglena uses chlorophyll to make its own food., An amoeba surrounds its food with pseudopods.

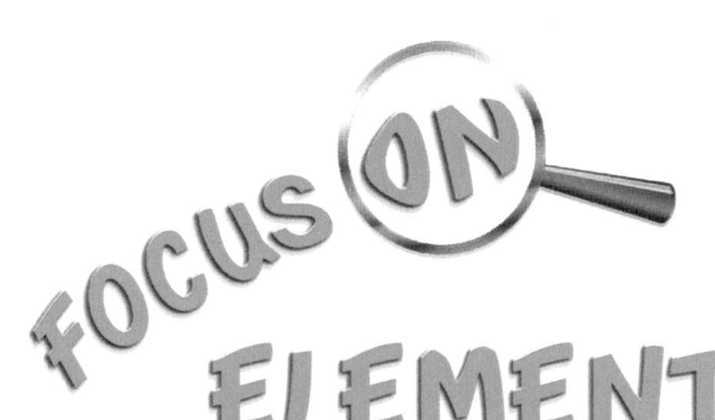

Grades K-4

ELEMENTARY

GRAPHICS

3rd Edition

LOOK MA–
NO HANDS!

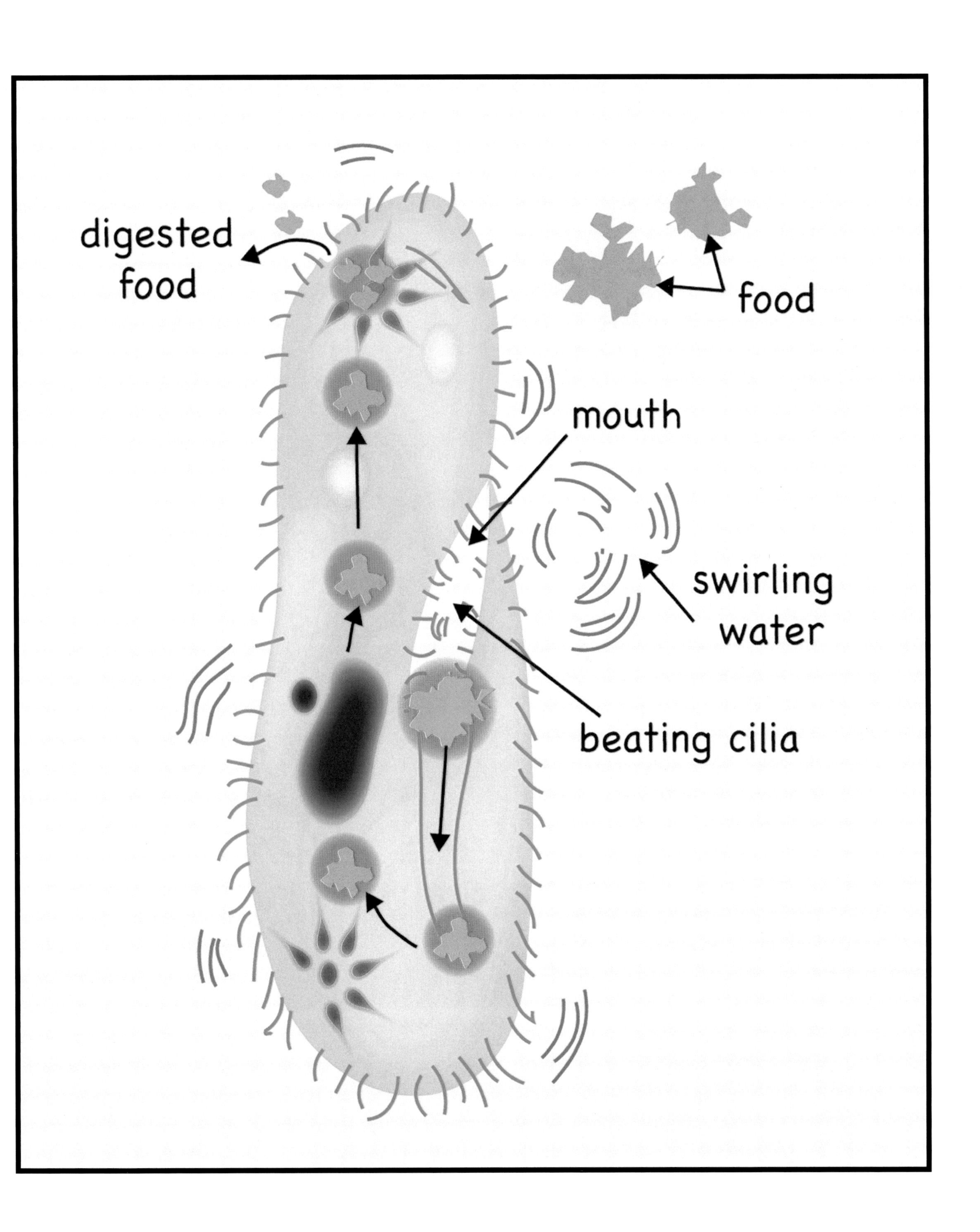
digested
food
food
mouth
swirling
water
beating cilia

More REAL SCIENCE-4-KIDS Books

by Rebecca W. Keller, PhD

Building Blocks Series yearlong study program — each Student Textbook has accompanying Laboratory Notebook, Teacher's Manual, Lesson Plan, Study Notebook, Quizzes, and Graphics Package

Exploring Science Book K (Activity Book)
Exploring Science Book 1
Exploring Science Book 2
Exploring Science Book 3
Exploring Science Book 4
Exploring Science Book 5
Exploring Science Book 6
Exploring Science Book 7
Exploring Science Book 8

Focus On Series unit study program — each title has a Student Textbook with accompanying Laboratory Notebook, Teacher's Manual, Lesson Plan, Study Notebook, Quizzes, and Graphics Package

Focus On Elementary Chemistry
Focus On Elementary Biology
Focus On Elementary Physics
Focus On Elementary Geology
Focus On Elementary Astronomy

Focus On Middle School Chemistry
Focus On Middle School Biology
Focus On Middle School Physics
Focus On Middle School Geology
Focus On Middle School Astronomy

Focus On High School Chemistry

Super Simple Science Experiments

21 Super Simple Chemistry Experiments
21 Super Simple Biology Experiments
21 Super Simple Physics Experiments
21 Super Simple Geology Experiments
21 Super Simple Astronomy Experiments
101 Super Simple Science Experiments

Note: A few titles may still be in production.

Gravitas Publications Inc.
www.gravitaspublications.com
www.realscience4kids.com

Made in the USA
Middletown, DE
25 November 2023